11+ Mathematics

Practice Papers 1

Hachette UK's policy is to use papers that are natural, renewable and recyclable products and made from wood grown in sustainable forests. The logging and manufacturing processes are expected to conform to the environmental regulations of the country of origin.

Orders: please contact Bookpoint Ltd, 130 Milton Park, Abingdon, Oxon OX14 4SB. Telephone: +44 (0)1235 827827. Lines are open 9.00a.m.–5.00p.m., Monday to Saturday, with a 24-hour message answering service. Visit our website at www.galorepark.co.uk for details of other revision guides for Common Entrance, examination papers and Galore Park publications.

ISBN: 978 1 471849 26 8

© Steve Hobbs 2016
First published in 2016 by
Galore Park Publishing Ltd
An Hachette UK Company
Carmelite House
50 Victoria Embankment
London EC4Y 0DZ
www.galorepark.co.uk
Impression number   10 9 8 7 6 5 4 3 2 1
Year   2020 2019 2018 2017 2016

All rights reserved. Apart from any use permitted under UK copyright law, no part of this publication may be reproduced or transmitted in any form or by any means, electronic or mechanical, including photocopying and recording, or held within any information storage and retrieval system, without permission in writing from the publisher or under licence from the Copyright Licensing Agency Limited. Further details of such licences (for reprographic reproduction) may be obtained from the Copyright Licensing Agency Limited, Saffron House, 6–10 Kirby Street, London EC1N 8TS.

Typeset in India

Printed in the UK

Illustrations by Integra Software Services, Ltd.

A catalogue record for this title is available from the British Library.

Name: _____

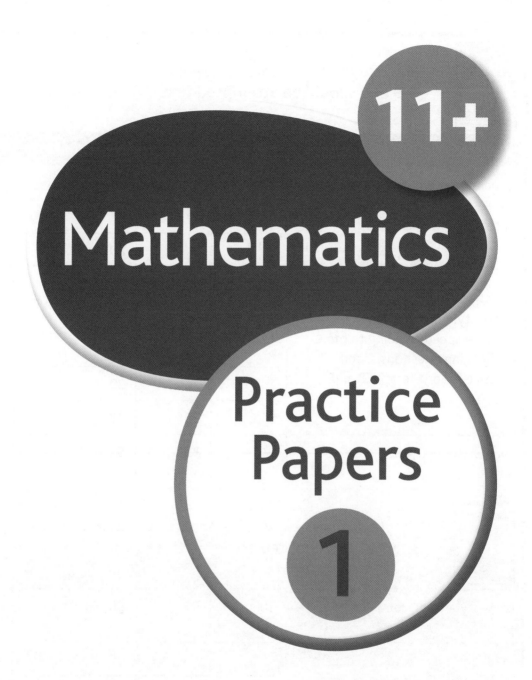

11+
Mathematics
Practice Papers
1

Steve Hobbs

Contents and progress record

How to use this book　　　　　　　6

Practice Papers 1 contains papers 1–9 and should be attempted first

Ideal for...	Paper	Page	Length (no. Qs)	Timing (mins)
General training for all 11+ and pre-tests, good for familiarisation of test conditions.	1	9	20	17:00
	2	11	8	17:00
	3	14	20	17:00
	4	16	8	20:00
Short-style tests designed to increase speed, building to more challenging questions. Good practice for pre-test, ISEB and CEM 11+.	5	19	20	14:00
	6	21	16	30:00
	7	26	30	15:00
	8	29	12	15:00
Long-style tests, good practice for GL.	9	33	50	50:00

Practice Papers 2 contains papers 10–13 and should be attempted after *Practice Papers 1*

	Paper	Page	Length	Timing
Long-style tests, good practice for GL.	10	9	50	50:00
Medium-length tests, building in difficulty. Good practice for pre-test, ISEB, CEM, GL or any independent school exam.	11	16	60	60:00
	12	22	70	90:00
	13	31	80	75:00

Answers　　　　　　　43

Speed	Question type	Score	Time
Slow	Multiple-choice	/ 29	:
Average	Multiple-choice	/ 36	:
Slow	Standard	/ 29	:
Average	Standard	/ 29	:
Fast	Standard	/ 29	:
Fast	Standard	/ 67	:
Fast	Standard	/ 54	:
Fast	Standard	/ 51	:
Average	Multiple-choice	/ 96	:
Realistic	Standard	/ 98	:
Average	Multiple-choice	/ 114	:
Realistic	Standard	/ 192	:
Fast	Standard	/ 174	:

How to use this book

Introduction

These *Practice Papers* have been written to provide final preparation for your 11+ Maths test.

Practice Papers 1 includes nine model papers with a total of 184 questions. There are …

- four training tests, which include some simpler questions and slower timing designed to develop confidence
- four tests in the style of pre-tests, ISEB (Independent Schools Examination Board) and short-format CEM (Centre for Evaluation and Monitoring)/bespoke tests in terms of difficulty, speed and question variation
- one test in the style of the longer format GL (Granada Learning)/bespoke tests in multiple-choice question format.

Practice Papers 2 includes four model papers with a total of 260 questions. These papers include …

- one in the style of a longer-format GL bespoke test in standard question format
- three further longer-format tests with challenging content, speed and question variation to support all 11+ tests.

Practice Papers 1 will help you …

- become familiar with the way long-format 11+ tests are presented
- build your confidence in answering the variety of questions set
- work with the most challenging questions set
- tackle questions presented in different ways
- build up your speed in answering questions to the timing expected in the 11+ tests.

Pre-test and the 11+ entrance exams

The Galore Park 11+ series is designed for pre-tests and 11+ entrance exams for admission into Independent Schools. These exams are often the same as those set by local Grammar Schools too. 11+ Maths tests now appear in different formats and lengths and it is likely that if you are applying for more than one school, you will encounter more than one of type of test. These include:

- pre-tests delivered on-screen
- 11+ entrance exams in different formats from GL, CEM and ISEB
- 11+ entrance exams created specifically for particular independent schools.

Tests are designed to vary from year to year. This means it is very difficult to predict the questions and structure that will come up, making the tests harder to revise for.

To give you the best chance of success in these assessments, Galore Park has worked with 11+ tutors, independent school teachers, test writers and specialist authors to create these *Practice Papers*. These books cover the styles of questions and the areas of Maths that typically occur in this wide range of tests.

Because 11+ tests now aim to include variations in the content and presentation of questions, making them increasingly difficult to revise for, **both** *books should be completed as essential preparation for all 11+ Maths tests.*

For parents

These *Practice Papers* have been written to help both you and your child prepare for both pre-test and 11+ entrance exams.

For your child to get maximum benefit from these tests, they should complete them in conditions as close as possible to those they will face in the exams, as described in the 'Working through the book' section below.

Timings get shorter as the book progresses to build up speed and confidence.

Some of these timings are very demanding and reviewing the tests again after completing the books (even though your child will have some familiarity with the questions) can be helpful, to demonstrate how their speed has improved through practice.

For teachers and tutors

This book has been written for teachers and tutors working with children preparing for both pre-test and 11+ entrance exams. The variations in length, format and range of questions is intended to prepare children for the increasingly unpredictable tests encountered, with a range of difficulty developed to prepare them for the most challenging and on-screen adaptable tests.

Working through the book

The **Contents and progress record** helps you to understand the purpose of each test and track your progress. Always read the notes in this record before beginning a test as this will give you an idea of how challenging the test will be!

You may find some of the questions hard, but don't worry. These tests are designed to build up your skills and speed. Agree with your parents on a good time to take the test and set a timer going. Prepare for each test as if you are actually going to sit your 11+ (see 'Test day tips' on page 8):

- Complete the test with a timer, in a quiet room, noting down how long it takes you, writing your answers in pencil. Even though timings are given, you should complete ALL the questions.
- Mark the test using the answers at the back of the book.
- Go through the test again with a friend or parent and talk about the difficult questions.
- Have another go at the questions you found difficult and read the answers carefully to find out what to look for next time.

The **Answers** are designed to be cut out so that you can mark your papers easily. Do not look at the answers until you have attempted a whole paper. Each answer has a full explanation so you can understand why you might have answered incorrectly and how to award the marks.

When you have finished a test, turn back to the Contents and progress record and fill in the boxes:

- Write your total number of marks in the 'Score' box
- Note the time you took to complete ALL the questions in the 'Time' box.

After completing both books you may want to go back to the earlier papers and have another go to see how much you have improved!

> **Test day tips**
> Take time to prepare yourself the day before you go for the test: remember to take sharpened pencils, an eraser and a watch to time yourself (if you are allowed – there is usually a clock present in the exam room in most schools). Take a bottle of water in with you, if this is allowed, as this will help to keep your brain hydrated and improve your concentration levels.
>
> … and don't forget to have breakfast before you go!

Continue your learning journey

When you've completed these *Practice Papers*, you can carry on your learning right up until exam day with the following resources.

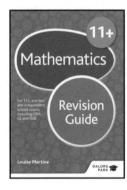 

The *Revision Guide* reviews all the areas of Maths you may encounter in your 11+ entrance exams as well as tips and guidance to excel above the other candidates. This includes additional material required by some schools that you may not have encountered.

Practice Papers 2 contains a further four model papers and answers to improve your accuracy, speed and ability to deal with a wide range of questions under pressure.

Paper 1

Test time: 17 minutes

Circle the correct answer for each question.

1. 550 − 337 (1)
   (a) 223 (b) 227 (c) 217 (d) 113 (e) 213

2. 134 + 259 (1)
   (a) 393 (b) 383 (c) 483 (d) 395 (e) 295

3. 38 × 4 (1)
   (a) 160 (b) 152 (c) 156 (d) 162 (e) 142

4. $\frac{3}{4} - \frac{1}{2}$ (2)
   (a) 1 (b) $\frac{1}{3}$ (c) $\frac{1}{2}$ (d) $\frac{1}{6}$ (e) $\frac{1}{4}$

5. 128 ÷ 4 (1)
   (a) 32 (b) 36 (c) 27 (d) 37 (e) 64

6. What is the product of 16 and 9? (1)
   (a) 140 (b) 134 (c) 148 (d) 154 (e) 144

7. What is the sum of 58 and 67 rounded to the nearest 10? (2)
   (a) 120 (b) 100 (c) 150 (d) 130 (e) 110

8. 673 + _____ = 945 (1)
   (a) 272 (b) 1618 (c) −272 (d) 238 (e) −238

9. Richard and Samir share a tin of 45 biscuits. They share them in the ratio 7:2 How many sweets will Richard get? (2)
   (a) 10 (b) 35 (c) 14 (d) 38 (e) 9

10. Estimate the number of circles in the rectangle. Do not count them. (1)

    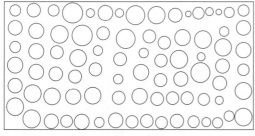

    (a) 50 (b) 90 (c) 160 (d) 200 (e) 130

11. The rectangle below consists of 28 squares. If $\frac{1}{4}$ of the shares were shaded, how many squares would be shaded? (1)

    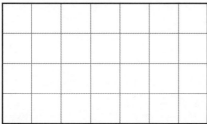

    (a) 4 (b) 7 (c) 8 (d) 9 (e) 14

Turn over to the next page

12 Jane worked for 7 hours delivering leaflets and her friend Sarah worked for 1 hour. In total they are paid £24
How much does Sarah get paid? (2)
(a) £3   (b) £1   (c) £6   (d) £21   (e) £4

13 Which fraction is equivalent to $\frac{6}{20}$? (1)
(a) $\frac{3}{5}$ (b) $\frac{5}{9}$ (c) $\frac{2}{3}$ (d) $\frac{5}{10}$ (e) $\frac{3}{10}$

14 Which decimal number is the same as the fraction $\frac{3}{4}$? (1)
(a) 0.34   (b) 0.75   (c) 0.075   (d) 0.7   (e) 0.0075

15 What is 750 centimetres converted into metres? (1)
(a) 7.5 m   (b) 750 m   (c) 0.750 m   (d) 0.075 m   (e) 75 m

16 The length of this rectangle is twice its width. What is the perimeter of the rectangle? (3)

Not drawn accurately

(a) 240 cm   (b) 260 cm   (c) 360 cm   (d) 380 cm   (e) 180 cm

17 What is the mean number of runs scored by these cricketers? (2)
Jamil   49 runs   Danny   76 runs   Diego   52 runs   Peter   27 runs
(a) 86 runs   (b) 51 runs   (c) 56 runs   (d) 72 runs   (e) 52 runs

18 What is the value of a in the following equation? (2)
$9 + a = 12$
(a) 21   (b) 12   (c) −3   (d) 3   (e) 9

19 What is the value of b in the following equation? (2)
$b - 7 = 2$
(a) 2   (b) −5   (c) −9   (d) 5   (e) 9

20 Which shape below is irregular? (1)

Record your results and move on to the next paper

Paper 2

Test time: 17 minutes

Circle the correct answer for each question.

1 Bob buys 5 cups of hot chocolate each costing £1.85 and 4 cakes each costing £1.60
   (a) How much will this cost? (3)
       (i) £15.65   (ii) £15.45   (iii) £11.75   (iv) £9.75   (v) £16.25
   (b) He pays with a £20 note. What is the fewest number of coins he could receive to get the correct change? (1)
       (i) 3 coins   (ii) 4 coins   (iii) 5 coins   (iv) 6 coins   (v) 7 coins
   (c) The next day there is a 'Buy one, get one free' offer on the cakes. He buys 5 cups of hot chocolate and 4 cakes and pays with a £20 note. How much change should he receive? (3)
       (i) £8.25   (ii) £9.35   (iii) £10.75   (iv) £6.55   (v) £7.55
   (d) What is your answer to part (c) rounded to the nearest £1.00? (1)
       (i) £7   (ii) £8   (iii) £9   (iv) £10   (v) £11
   (e) Another customer receives £1.63 in change. Which combination of coins is this? (1)
       (i) £1, 50p, 20p, 1p   (ii) £1, 50p, 10p, 2p, 1p
       (iii) £2, 50p, 10p, 2p, 1p   (iv) £1, 50p, 10p, 5p   (v) £1, 10p, 2p, 1p

2 Here are some long jump distances.
   3.65 m   3.45 m   3.25 m   3.57 m
   (a) What is the mean of these long jump distances? (2)
       (i) 3.50 m   (ii) 13.92 m   (iii) 3.55 m   (iv) 3.45 m   (v) 3.48 m
   (b) The longest jump is 3.65 m and the smallest jump is 3.25 m. What is the range of these distances? (1)
       (i) 3.65 m   (ii) 3.25 m   (iii) 13.92 m   (iv) 0.4 m   (v) 0.45 m

3 Peter, Alex and Michelle share sweets in a bag in the ratio 4:3:2
   Michelle receives 8 sweets.
   (a) How many sweets does Peter receive? (2)
       (i) 4   (ii) 8   (iii) 9   (iv) 12   (v) 16
   (b) How many sweets are in the bag at the start? (2)
       (i) 9   (ii) 24   (iii) 36   (iv) 48   (v) 72

Turn over to the next page

4 Which of these images shows the net of a cube? (1)

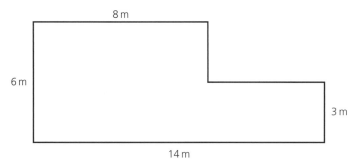

5 Below is a plan view of Rebecca's bedroom.

8 m
6 m
3 m
14 m

Not drawn accurately

(a) What is the perimeter of Rebecca's bedroom? (1)
   (i) 40 m   (ii) 44 m   (iii) 48 m   (iv) 80 m   (v) 84 m

(b) What is the area of Rebecca's bedroom? (3)
   (i) 48 m²   (ii) 60 m²   (iii) 66 m²   (iv) 84 m²   (v) 90 m²

(c) Carpet costs £8.40 per square metre. How much will it cost to carpet Rebecca's bedroom? (1)
   (i) £600.50   (ii) £582.60   (iii) £560.60   (iv) £550.50   (v) £554.40

6 Which number is missing from each number sentence?

(a) 26 − 13 = 20 − ____ (2)
   (i) 6   (ii) 7   (iii) 4   (iv) 8   (v) 5

(b) 50 ÷ ____ = 2 × 5 (2)
   (i) 10   (ii) 2   (iii) 15   (iv) 5   (v) 50

7 The tally chart shows how children travelled to a local primary school on the first day of term.

Type of travel	Tally
Walking	|||| |||| |||| |||| |||| |||| |||| |||| |||| |||| |||| ||
Car	|||| |||| |||| |||| |||| |||| |||| |||| |||| |||| |||| |||| |||| |||
Bicycle	|||| |||| |||| |||| |||| |||| ||
Bus	|||| |||| |||| |||| |||| |||| ||
Scooter	|||| ||||	

(a) How many children travelled by bicycle? (1)
   (i) 5   (ii) 15   (iii) 32   (iv) 26   (v) 52

(b) How many children went to school on this day? (2)
   (i) 120   (ii) 132   (iii) 134   (iv) 136   (v) 200
(c) How many more children came by car than by bus? (2)
   (i) 16   (ii) 24   (iii) 36   (iv) 70   (v) 72
(d) What fraction of the children walked, cycled or scooted to school? (3)
   (i) $\frac{1}{2}$ (ii) $\frac{1}{4}$ (iii) $\frac{3}{25}$ (iv) $\frac{9}{50}$ (v) $\frac{1}{75}$

8 One side of a square is drawn on the grid below. If the square fits on the grid, what are the co-ordinates of the missing vertices. (2)

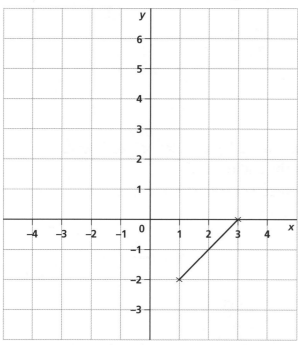

(a) (2, 1) (0, −1)
(b) (1, 2) (−1, 0)
(c) (−1, 2) (0, −1)
(d) (2, 2) (−1, 1)
(e) (1, 1) (−1, −1)

Record your results and move on to the next paper

Score ☐ / 36   Time ☐ : ☐

Paper 3

Test time: 17 minutes

1. Write the year represented by the Roman numerals MLIX. _____ (1)
2. −18 + 27 = _____ (1)
3. Insert the missing digits in this addition. (2)

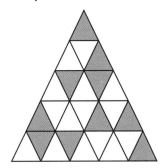

4. 138 ÷ 6 = _____ (1)
5. 1254 × 3 = _____ (1)
6. There are 5272 seats in the Royal Albert Hall. During the interval, 2346 of the audience leave the auditorium. How many stay in the auditorium? _____ (1)
7. What number between 130 and 140 is exactly divisible by 12? _____ (1)
8. What is the product of the first three prime numbers? _____ (1)
9. A television costs £360
   In the sale it is reduced by 15%. What is the sale price of the television?
   _____ (2)
10. What fraction of the shape below is shaded grey? Give your answer in its simplest form. _____ (2)

11. $5x + 3y = 43$
    Find the value of y when $x = 5$
    $y =$ _____ (3)
12. What is the perimeter of this rectangle? (2)

Not drawn accurately

13 What is the area of this shape? (1)

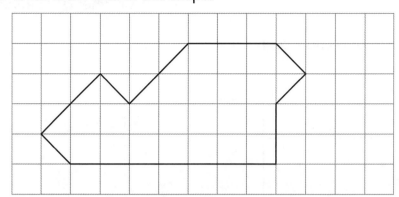

14 A postman has four parcels to deliver. The mass of the parcels are 3.24 kg, 0.81 kg, 4.22 kg and 1.7 kg. What is the total mass of the parcels? _____ (1)

15 $\frac{3}{4}$ of 64 = _____ (2)

16 What is the value of $3y + 8$ when $y = 9$? _____ (1)

17 The finishing times of the sprinters in a 100-metre race are given below.

   9.81   9.85   9.91   9.83   10.0

What is the mean time? _____ (2)

18 On a map, 1 cm represents 5 km. What is the actual distance represented by a line of 9 cm on the map? _____ (1)

19 Find the missing angle in the quadrilateral below. y = _____ (2)

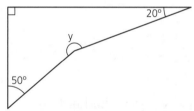

Not drawn accurately

20 A scale model car is 40 cm long. The scale is 1 : 10
How long is the actual car? _____ (1)

Record your results and move on to the next paper

Paper 4

Test time: 20 minutes

1 Below is a plan of two classrooms and a connecting corridor.

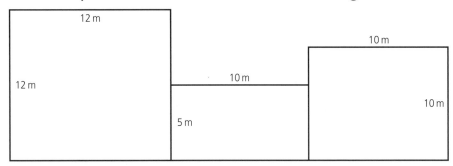

Not drawn accurately

(a) What is the perimeter of the larger classroom? _____ (1)
(b) What is the difference in area between the two classrooms? _____ (3)
(c) During the summer holidays, the caretaker tiles the classroom floors. The tiles are 25 cm × 25 cm in size. How many tiles will the caretaker need to tile the larger classroom? _____ (2)
(d) Tiles cost 10p each. How much will it cost to tile the larger classroom?
_____ (1)
(e) What is the difference in cost between tiling the two classrooms?
_____ (3)

2 Twickenham Stadium has 82 000 seats.
Half an hour before the start of a match, 13 484 seats were empty.
(a) How many people were in their seats? _____ (1)
(b) If only 25% of the empty seats had been sold, how many people had not yet entered the stadium? _____ (1)
(c) 8000 children watched the game. The ratio of boys to girls was 3 : 1

How many boys watched the game? _____ (2)

3 Write in words the number shown on the abacus below.

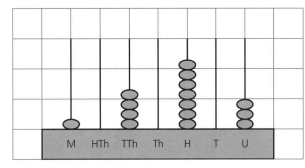

_____ (1)

4 Write the following amounts of money in order, smallest first. (1)

£8.67　　　　£8.76　　　　£6.87　　　　£6.78　　　　£7.86

____　　　　____　　　　____　　　　____　　　　____

5 Azsvin asked members of his youth club about the sports they play. His results are shown in the Venn diagram below.

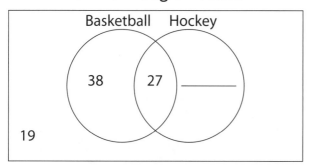

(a) How many children play both basketball and hockey? _____ (1)

(b) Altogether, 48 children play hockey. How many children play only hockey and no other sport(s)? _____ (1)

(c) How many children took part in the survey? _____ (1)

(d) How many children play basketball? _____ (1)

(e) How many children play just one sport? _____ (1)

6 Paul was thinking of a number, x. When he subtracted 5 and then multiplied by 3 the answer was 12

(a) Write down an expression to help find Paul's number. _____ (2)

(b) What number was Paul thinking of? _____ (2)

7 Mandy is decorating her bathroom. She has tiled a quarter of one wall, as shown in the diagram. She wants to reflect the pattern along the horizontal and vertical axes marked on the diagram. Colour squares to show the pattern on the finished wall. (2)

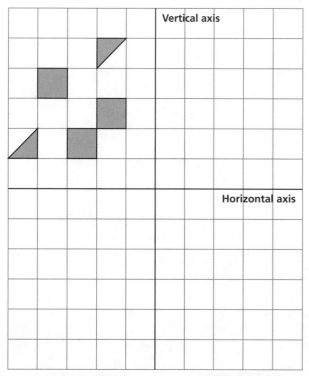

Turn over to the next page

8 The Great Pyramid in Egypt is a square-based pyramid.
   (a) Draw the net of a square-based pyramid below. (1)

   (b) The base of the Pyramid of Khufu at Giza has sides of 230 m. Lawrence wants to walk around the base of this pyramid. How far is this? _____ (1)

Record your results and move on to the next paper

Score ☐ / 29   Time ☐ : ☐

Paper 5

Test time: 14 minutes

1 Put these lengths in order, starting with the smallest. (2)

$\frac{1}{2}$ m   51 cm   1 m   15 cm   49 cm   98 cm

_____ _____ _____ _____ _____ _____

2 891 + 569 = _____ (1)

3 923 − 394 = _____ (1)

4 Below is part of a 100 square.

4	5	6
14	15	16
24	25	26

What is the only prime number in this part of the 100 square? _____ (1)

5 Write down all the factor pairs of 24 (1)

(____, ____)   (____, ____)   (____, ____)   (____, ____)

6 $3^2 + 2^3$ = _____ (3)

7 1440 ÷ 6 = _____ (1)

8 62 of the 186 children on a school trip are boys. Write this as a fraction in its simplest form (lowest terms)? _____ (2)

9 $\frac{1}{2} \times \frac{4}{5}$ = _____ (1)

10 What is $\frac{3}{5}$ of 45 miles? _____ (2)

11 0.685 − 0.394 = _____ (1)

12 Jeremy and Lewis are going on a 78 mile car journey. They plan to share the driving in the ratio of 1 : 5

How many miles will Lewis drive? _____ (2)

13 What proportion of the rectangle is shaded? Give your answer as a fraction. _____ (1)

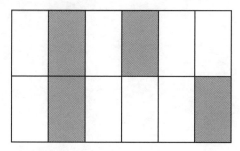

14 $3h + y = 29$

Find the value of y when $h = 7$

y = _____ (2)

15 What is the 10th term (number) in the sequence 3, 7, 11, 15 …? _____ (1)

Turn over to the next page

16 A building is in the shape of a regular pentagon. The length of one of its sides is 281 metres. What is the perimeter of this building? _____ (1)

17 Yazmin ran 4250 metres in the school track race. Write this distance in kilometres. _____ (1)

18 Namish's book is 8 inches long.
1 inch = 2.5 cm
Write the length of the book in centimetres. _____ (1)

19 Calculate the size of the missing angle in this triangle. _____ (2)

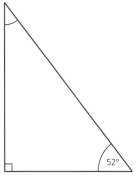

Not drawn accurately

20 The pie chart shows the favourite sports of children in a class.
If there are 32 children in the class, how many chose rugby as their favourite sport? _____ (2)

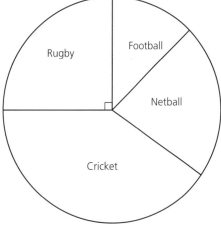

Record your results and move on to the next paper

Score ☐ / 29   Time ☐ : ☐

Paper 6

Test time: 30 minutes

1 The chart shows the distances, in kilometres, between some French cities.

Dijon					
194	Lyon				
503	313	Marseille			
639	685	985	Nantes		
315	466	775	385	Paris	
622	740	1050	110	353	Rennes

(a) Pierre wants to travel from Lyon to Rennes. How far is this? _____ (1)

(b) What is the distance between Nantes and the next closest city. _____ (1)

(c) What is the difference in the distance between Marseille and Rennes and Marseille and Paris? _____ (3)

2 Freddie ran for 3 hours at an average speed of 12 km/h and for a further 20 minutes at 9 km/h.

(a) What total distance did Freddie cover? _____ (3)

(b) What was Freddie's average speed for his run? _____ (3)

3 The chart shows some test results from a class spelling test.

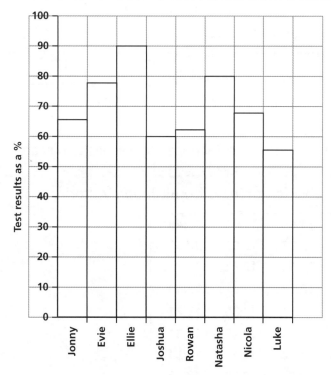

Turn over to the next page

(a) What was the average percentage score in the test? _____ (2)

(b) The test contained 50 spellings. How many more spellings did Natasha get right than Luke? _____ (3)

(c) What is the range of spelling scores for these eight children? _____ (2)

4 Jensen's car travels 9 km on a litre of fuel.

(a) How much fuel will he need for a journey of 450 km? _____ (1)

(b) Fuel costs £1.10 per litre. How much will the fuel for this journey cost? _____ (1)

5 A cube with sides of 5 cm is made up of 1 cm cubes.

5 cm
5 cm
5 cm

Not drawn accurately

(a) How many 1 cm cubes are there altogether? _____ (1)

(b) How many of the cubes have none of their faces showing? _____ (2)

(c) What percentage of the cubes has at least one face showing? _____ (1)

6 Michael buys 7 silk ties costing £8.95 each and 3 shirts costing £11.99 each.

(a) How much is this altogether? _____ (3)

(b) If he spends £100 or more, he gets £20 off the total price. He decides to buy an extra tie. How much does he pay altogether now? _____ (2)

7 Thomas is looking at the train timetable for journeys from London Victoria.

London Victoria	10:06	10:34	11:55	12:01	12:48
Clapham Junction	10:13	10:41	12:02	12:09	12:55
East Croydon	10:28	10:52	12:13	12:21	13:07
Purley	10:34		12:23		13:14
Redhill	10:41		12:29	12:35	13:21
Gatwick	10:55	11:08	12:39	12:45	13:31
Three Bridges	10:58	11:11	12:42	12:48	13:34
Haywards Heath	11:07	11:20	12:52	12:58	13:43
Brighton	11:25	11:38	13:11	13:18	14:01

(a) How long will it take him to get to from London Victoria to Brighton if he catches the 10:34 train? _____ (1)

(b) If Thomas needs to be at Gatwick for 12:30, which train should he catch? _____ (1)

(c) The 11:55 train from London Victoria arrives at Haywards Heath 9 minutes late. At what time does it arrive? _____ (1)

(d) Gordon arrives at East Croydon station at 12:25 and catches the next train to Brighton. How long after he arrives at East Croydon station does he arrive in Brighton? _____ (1)

(e) 536 000 passengers arrive in London by train between 07:00 and 10:00 each weekday morning. During the whole day, 981 000 arrive in London by train. How many people arrive in London by train outside of the hours of 07:00 to 10:00? _____ (1)

8 Mr Green recorded the types of trees in the park. The pictogram below shows his results.

Oak	🌳 🌳 🌳 🌳
Ash	🌳 🌳 🌳 🌳 🌳
Horse chestnut	🌳 🌳 🌳
Hazel	🌳 🌳 🌳 🌳 🌲
Yew	🌳 🌲
Sycamore	🌳 🌳 🌳 🌳 🌲
Hawthorn	🌳 🌳 🌳 🌳 🌳

Key: 🌳 respresents 2 trees

(a) How many ash trees are there in the park? _____ (1)

(b) How many trees are there in the park altogether? _____ (2)

(c) Mr Green finds out that the oak trees grow to a maximum height of 40 metres, the ash trees to a maximum height of 35 metres and the horse chestnut trees to a maximum height of 28 metres. If each of these trees in the park grows to its maximum height, what is the total combined height of these trees, in metres? _____ (4)

(d) The park authority wants the ratio of horse chestnut trees to other trees in the park to be 1 : 10

How many more trees will need to be planted? _____ (2)

(e) In a storm, for every 2 trees blown down, 3 are left standing. How many trees are left standing? _____ (2)

Turn over to the next page

9 The list below shows the number of goals Romelda has scored in her last 24 netball matches.

| 7 | 6 | 6 | 8 | 4 | 3 | 9 | 5 | 4 | 7 | 2 | 7 |
| 4 | 6 | 5 | 3 | 5 | 7 | 6 | 5 | 8 | 9 | 3 | 6 |

(a) Complete the tally chart below. (1)

Number of goals	Tally of goals scored	Frequency
2		
3		
4		
5		
6		
7		
8		
9		

(b) What is Romelda's mean number of goals? _____ (2)

(c) What is Romelda's modal number of goals? _____ (1)

(d) What is Romelda's median number of goals? _____ (2)

(e) How many goals would Romelda need to score in her next match to increase her mean score to 6? _____ (2)

10 Mr Reeve is taking 9 children on the ski trip. They have paid a total of £8901 and each child has paid the same amount.

How much has each child paid? _____ (1)

11 Percy's garden is 13 m². He draws a plan of his garden and splits it into metre squares as shown on the co-ordinate grid below.

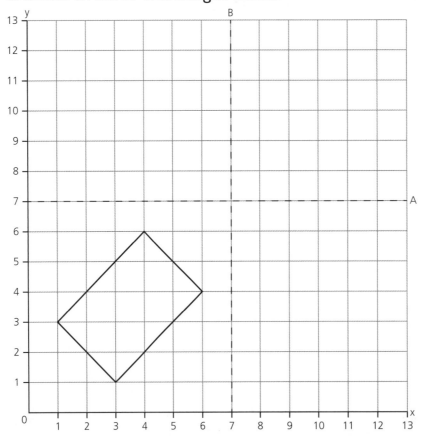

(a) What are the co-ordinates of the vertices of the large rectangular paving stone shown? (1)

(____, ____) (____, ____) (____, ____) (____, ____)

(b) Percy wants to lay three more paving stones in his garden. He works out the position of the next paving stone by reflecting the original paving stone in the dashed line labelled 'A'. Mark the position of this stone on the co-ordinate grid. (1)

(c) To find the position of the other two paving stones, reflect both the existing paving stones in the dashed line labelled 'B'. Mark the position of the two new stones on the co-ordinate grid. (2)

12 Andrew is thinking of a number, x. When he multiplies x by 9 and then halves the result, the answer is 27

What is Andrew's number, x? _____ (2)

13 Monty plants 25 rows of potatoes with 48 potato plants in each row.

(a) How many potato plants is this altogether? _____ (1)

(b) If each plant produces 2 kg of potatoes, what is the total mass of potatoes from Monty's plants? _____ (1)

14 Mr Wicker and his family travelled 3936 km across the USA from Los Angeles to New York. The journey took 24 days and they travelled for 8 hours each day. How far, on average, did they travel during each hour they were travelling? _____ (2)

15 Look at the pattern below.

Pattern   1   2   3

Pattern 1 uses 3 straws, pattern 2 uses 5 straws, ...
How many straws would be needed to make the 9th pattern in the sequence? _____ (2)

16 Look at the triangle on the grid.

The original triangle is translated three squares to the right and one square upwards. Mark the position of the resulting triangle on the grid. (1)

Record your results and move on to the next paper

Score ☐ / 67   Time ☐ : ☐

Paper 7

Test time: 15 minutes

1. Write the correct sign (<, > or =) in the number sentence below. (2)
   6 + (23 × 5) _____ (11 × 11) − 3

2. Write the next three numbers in the number sequence below. (3)
   2, 5, 11, 23, _____, _____, _____

3. What is the highest common factor of 24 and 36? _____ (2)

4. What are the prime factors of 42? _____ (1)

5. $2\frac{1}{2}$ m = _____ mm (1)

6. Write the following numbers in order of size, starting with the smallest. (1)
   67 666   767 667   666 766   776 677   76 776

7. Robert has read $\frac{2}{3}$ of a book. He has 166 pages more to read. How many pages are in the book? _____ (1)

8. What is the volume of a cuboid measuring 3.5 cm by 7 cm by 5 cm? _____ (1)

9. On the number grid, shade all the multiples of 4 and circle all the multiples of 6 (2)

33	34	35	36	37	38
43	44	45	46	47	48
53	54	55	56	57	58
63	64	65	66	67	68

10. Three consecutive numbers have a sum of 18 and a product of 210
    What are the three numbers? _____ (2)

11. $\frac{4}{5} \div 2 =$ _____ (2)

12. $\frac{1}{5} \times \frac{1}{3} =$ _____ (1)

13. $4\frac{5}{6} - 2\frac{1}{6} =$ _____ (2)

14. $3\frac{2}{6} + 2\frac{3}{6} =$ _____ (2)

15. What is 15% of £660? _____ (2)

16. Jasmine, Chloe and Freddie are completing a sponsorship event to raise money for a local charity. They raise amounts of money in the ratio 3 : 5 : 4
    They raise £1068 altogether. How much does Chloe raise? _____ (3)

17. Suzanna is thinking of a two-digit number. She gives the following clues:
    It is a prime number less than 30
    It has two digits.
    The sum of the digits is 5
    What is Suzanna's number? _____ (2)

18. Small cubes have length 2 cm, width 2 cm and height 2 cm. How many of these small cubes will fit inside a large cube with length 20 cm, width 20 cm and height 20 cm?
    _____ (2)

19 The diagram shows a three-dimensional shape. Name the shape and complete the information about the number of faces, edges and vertices it has. (4)

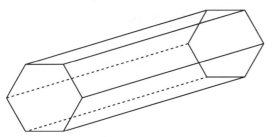

Name of shape: _____ Number of faces: _____
Number of edges: _____ Number of vertices: _____

20 The circle below has a radius of 6 cm. What is the circumference (perimeter) of the circle?
(Circumference = 3.14 × diameter) _____ (1)

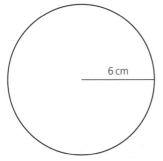

Not drawn accurately

21 A number (n) is divided by 4 and 6 is added to the result. Write an expression in terms of n for the final answer. _____ (1)

22 Work out 346 × 3 using any written method. _____ (1)

23 Work out 8736 divided by 6 using any written method. _____ (1)

24 The mass of some babies at birth is as follows.
2.9 kg   3.7 kg   3.1 kg   3.9 kg   4.2 kg   3.6 kg   3.1 kg   4.1 kg
What is the mean weight? _____ (2)

25 Work out the value of a in the equation below.
$4(5 + 2a^2) = 52$
$a =$ _____ (4)

Turn over to the next page

26 What is the ratio of apples to oranges if there are 120 apples and 90 oranges?
_____ (1)

27 Two people are decorating a house. One works for 2 hours and the other for 3 hours. They are paid a total of £20
The money is shared in the ratio of the number of hours worked. How much does each person receive? _____ (4)

28 On the axes below, plot the line $y = 2x$. (1)

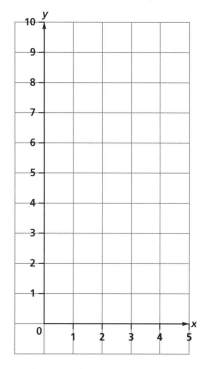

29 2.5 cm = 1 inch
22.5 cm = _____ inches (1)

30 Express 134 minutes in hours and minutes. _____ hours _____ minutes (1)

Record your results and move on to the next paper

Score ☐ / 54 Time ☐ : ☐

Paper 8

Test time: 15 minutes

1 The children in three Year 6 classes are given a choice of after-school clubs. They can choose to join the swimming club, the running club or the rugby club.
Their choices are shown in the table below.

	Swimming club	Running club	Rugby club	Total
Mr Richards' class	12	8		28
Mrs Flynn's class		13	7	30
Mrs Thomas's class	5		9	
Total				85

(a) Complete the table. (4)

(b) What percentage of the children in Mrs Flynn's class chose the swimming club?
_____ (2)

(c) Last Tuesday, the ratio of children from the swimming club who forgot their swimming kit to those that didn't was 1:2
How many children forgot their swimming kit? _____ (2)

(d) Write down the ratio of children that chose swimming : running : rugby in Mr Richard's class. Remember to write the ratio in its simplest form.
_____ (1)

(e) A new table tennis club is set up and 20% of the children join this club instead. How many children is this? _____ (2)

2 Toby has eight cards showing different numbers, like the cards below.

On the probability scales below, 0 means that an outcome is impossible and 1 means that it is certain. Mark the probability of each outcome on the diagram with an X.

(a) Toby picks a card that shows a square number. (1)

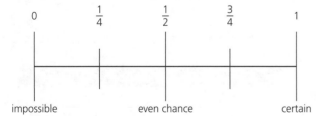

Turn over to the next page

(b) Toby picks a card that shows an even number. (1)

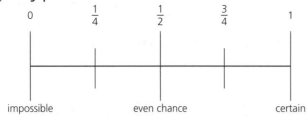

(c) Toby picks a card that shows a factor of 24 (1)

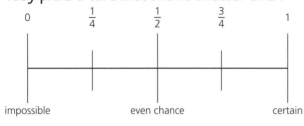

3 Draw the image of the shape when it is reflected in line A. (1)

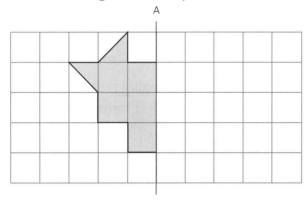

4 Draw the image of the shape when it is rotated 90° clockwise about point A. (1)

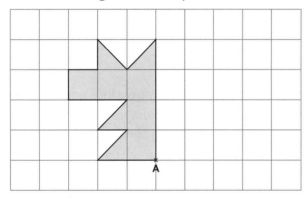

5 Look at the information in the table below.

Name	Hair colour	Height (cm)	Shoe size	School sweater colour	Wear a tie
Anne	black	151	7	black	Yes
Joe	blonde	145	6	blue	No
Darren	brown	153	7	blue	Yes
Arjun	brown	150	8	red	No
Ben	ginger	141	7	black	Yes
Phoebe	blonde	143	5	green	No

(a) How many of the children have brown hair and wear a red sweater?

_____ (1)

(b) How many of the children wear shoes smaller than size 7? _____ (1)

(c) How many children are taller than Joe? _____ (1)

(d) How many children have a sweater the same colour as their hair? _____

(1)

6 The isosceles triangle on the grid is translated two squares to the left. What are the co-ordinates of the three vertices? _____ (3)

7 Alice went on holiday to *Unbelievableland* where the unit of currency is the Bling (B). The exchange rate at her bank was £1 = 4 Blings.

(a) Alice exchanged £350

How many Blings did she receive? _____ (1)

(b) Alice bought some suntan cream for 120 B, a hat for 85 B, a guide book for 100 B and a flag for 230 B. How much was the total in UK pounds? _____ (2)

(c) Alice uses up all of her Blings so she changes some more money at her hotel. She changes £120 at an exchange rate of £1 = 3.5 Blings. How many more Blings would she have received if she had exchanged the money at her bank before she went on holiday? _____ (3)

(d) At the end of her holiday, Alice still has 150 Blings. She changes these back into pounds at a rate of 3 Blings = £1

How many UK pounds does she receive? _____ (1)

(e) Alice sees the same suntan cream she bought in *Unbelievableland* being sold for £28.50

Using the exchange rate 1 UK pound = 4 Blings, in which country was the suntan cream cheaper and by how many Blings? _____ (2)

Turn over to the next page

8 Beatrice has written the first 100 pages of her new book. On average, there are 350 words on each page.
   (a) How many words has she written so far? _____ (1)
   (b) Beatrice writes 5 more pages to finish her book. How many words are in the book now? _____ (2)
   (c) Beatrice wrote 3 pages of her book each day. How many days did it take her to write the book? _____ (2)

9 Motorcycle Mike jumps over parked cars as part of his stunt show. The cars are parked side by side with no gaps. Each of the six cars is 1750 mm wide.
   (a) What distance does Motorcycle Mike jump in metres? _____ (2)
   (b) Mike decides to increase the distance of his jump by parking the cars end to end. Each car is 4740 mm long. What distance is Motorcycle Mike now jumping? Give your answer in metres. _____ (2)

10 Eight runners run the London Marathon and raise a total of £17 248
   (a) How much money, on average, does each runner raise? Give your answer to the nearest £10 _____ (2)
   (b) There are 26 full miles in a marathon. How much money, on average, does each runner raise per mile? Give your answer to the nearest pound. _____ (2)

11 Alan likes all fruit, but in different amounts. He thinks an apple is worth 2 bananas and a banana is worth 4 cherries. He has these equations to show his preferences.
   $a = 2b$ and $b = 4c$ (where a = apples, b = bananas and c = cherries)
   (a) How many cherries is one apple worth? _____ (2)
   (b) A kiwi fruit is worth 3 apples. How many bananas is a kiwi worth? _____ (2)

12 Jo has written down a number sequence, but cannot read the sixth number.
   1,   4,   13,   40,   121,

   What should the sixth number be? _____ (2)

Record your results and move on to the next paper

Score ☐ / 51   Time ☐ : ☐

Paper 9

Test time: 50 minutes

Circle the correct answer for each question.

1 What number should go in the box? (2)

   14 + 14 + 14 + 14 + 14 + 14 = ☐ × 12

   (a) 14   (b) 84   (c) 12   (d) 7   (e) 16

2 Philippa has a party with 11 friends. They share 2 birthday cakes equally among themselves. What fraction of the birthday cakes does each person receive? (2)

   (a) $\frac{2}{11}$ (b) $\frac{1}{5}$ (c) $\frac{1}{6}$ (d) $\frac{11}{12}$ (e) $\frac{1}{12}$

3 Which of these statements is correct? (1)

   (a) $9.6 > 9\frac{3}{5}$ (b) $9.6 < 9\frac{1}{2}$ (c) $9.6 = 9\frac{1}{2}$ (d) $9.6 = 9\frac{3}{5}$ (e) $9.6 < 9\frac{2}{5}$

4 The profit made by Flying Engines Plc was four hundred and eighteen thousand and fifty-eight pounds. What is the amount in figures? (1)

   (a) £418 508   (b) £418 058   (c) £408 508   (d) £400 188   (e) £136 418

5 A local pet shop sells tropical fish. A group of students counted the fish and plotted a bar chart to show the results.

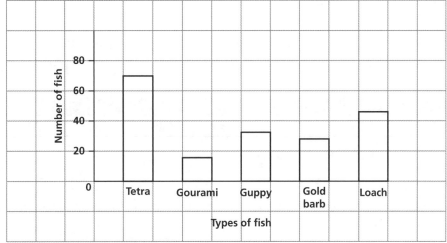

   How many more tetras are there than gold barbs? (2)

   (a) 40   (b) 35   (c) 30   (d) 45   (e) 25

6 The analogue clock shows the time in the afternoon. What is the equivalent digital time? (1)

   (a) 03:25   (b) 02:25   (c) 14:45   (d) 14:35   (e) 15:35

Turn over to the next page

7 What number is marked by the arrow on the scale below? (1)

   (a) 23.4   (b) 23.3   (c) 23.35   (d) 23.45   (e) 22.9

8 952 × 32 = 30 464

   What is 952 × 16? (2)

   (a) 7876   (b) 15 464   (c) 15 352   (d) 15 928   (e) 15 232

9 A 2-litre bottle of double strength blackcurrant squash is mixed with 6 times as much water to make diluted squash.

   How many 250-millilitre glasses can be filled with the diluted squash? (4)

   (a) 56   (b) 48   (c) 32   (d) 64   (e) 24

10 What is 3 − 1.7? (1)

   (a) 2.2   (b) 1.3   (c) 1.2   (d) 2.3   (e) 1.7

11 What shape do these clues describe? (1)
   - *I have two pairs of parallel sides.*
   - *My opposite sides are equal in length.*
   - *I do not have any right angles.*

   (a) a square   (b) a rectangle   (c) a trapezium   (d) a kite   (e) a parallelogram

12 The table below shows the number of conkers Henry collected in the local park during one week.

Day	Monday	Tuesday	Wednesday	Thursday	Friday	Saturday	Sunday
Number of conkers	32	43	37	33	26	26	20

   What is the mean number of conkers he collected? (2)

   (a) 31   (b) 43   (c) 32   (d) 26   (e) 20

13 The triangle is rotated 90° clockwise about the co-ordinates (4, 4). What are the new co-ordinates of point k? (1)

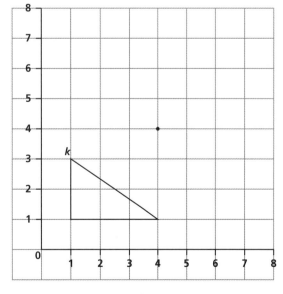

   (a) (5, 1)   (b) (7, 5)   (c) (3, 7)   (d) (1, 4)   (e) (3, 4)

14 Neil used the function machine below.

Input → ×8 → −2 → ÷2 → Output

If he used W as his input, what was his output? (3)

(a) $8W - 2$ (b) $8W - 1$ (c) $4W - 2$ (d) $4W - 1$ (e) $8W - \frac{1}{2}$

15 A square has co-ordinates (3, 3), (7, 3), (3, 7) and (7, 7). What are the co-ordinates of the middle point of the square? (2)

(a) (4, 3) (b) (5, 6) (c) (5, 5) (d) (6, 6) (e) (3, 6)

16 A mouse starts at (5, 3) on the grid below. He follows the lines of the grid and moves 3 squares west and then 6 squares south. Where does he stop? (2)

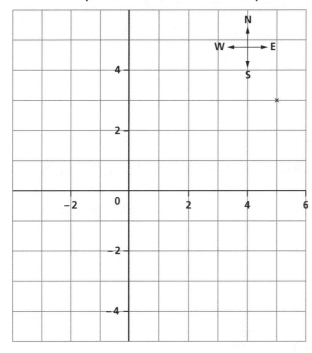

(a) (−2, −3) (b) (−3, 2) (c) (2, −3) (d) (0, −1) (e) (0, 2)

17 What are the missing digits in this addition calculation? (3)

9 __ + __ __ 3 = 3 7 5

(a) 2, 3, 7 (b) 2, 2, 7 (c) 3, 2, 8 (d) 2, 2, 8 (e) 3, 2, 7

18 What is 5280 ÷ 4? (1)

(a) 1200 (b) 1320 (c) 1070 (d) 1700 (e) 1120

19 What is 2.5 × (3.5 − 1.5) × 3? (2)

(a) 7.5 (b) 37.5 (c) 12 (d) 15 (e) 30.5

20 In a class of 25 children, 3 arrive late. What percentage of the children is this? (2)

(a) 3% (b) 30% (c) 12% (d) 6% (e) 20%

21 The numbers in this sequence increase by the same amount each time.

168, 174, 180, 186, 192, ...

Which one of these numbers will appear in this sequence? (2)

(a) 468 (b) 560 (c) 586 (d) 704 (e) 803

Turn over to the next page

22 Mr Panther has forgotten the code for his safe. He knows that:
- The code is a three-digit number below 200
- The number appears in the 8 times table.
- The three digits add up to 13

Which of these is the code for the safe? (3)

(a) 175   (b) 139   (c) 168   (d) 184   (e) 256

23 What is the cost of 24 mugs each costing 99p? (3)

(a) £24.24   (b) £36.24   (c) £23.76   (d) £24.76   (e) £20.76

24 Which child is closest to 15 months old on 15 October 2015? (2)

Child	Date of birth
Peter	11/08/14
Richard	21/07/14
Sean	20/11/14
Sally	26/10/14
Sara	05/02/15

(a) Peter   (b) Richard   (c) Sean   (d) Sally   (e) Sara

25 There are a total of 968 people at Royal Park School. There are 23 male teachers and 25 female teachers. The rest of the people are students. There are 46 more male students than female students. How many male students are there at Royal Park School? (5)

(a) 483   (b) 484   (c) 460   (d) 437   (e) 506

26 Below is part of the bus timetable for the route from Marton to Hopesfield.

Marton	06:10	07:15	07:40	07:50	08:15
Rickington	06:35	07:43	08:11	08:15	08:45
Barton	06:49	08:01	–	08:37	09:10
Brilliton	06:58	08:14	–	08:55	09:30
Chapplesmith	07:13	08:32	–	09:16	09:55
Hopesfield	07:26	08:49	08:59	09:35	10:15

It takes 5 minutes to walk from the bus stop to the town centre in Hopesfield. At what time should Tanya catch the bus from Marton if she wants to be in Hopesfield Town Centre at 9 o'clock? (2)

(a) 06:10   (b) 07:15   (c) 07:40   (d) 07:50   (e) 08:15

27 What temperature is shown on this thermometer? (1)

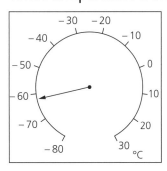

(a) −61 °C   (b) −65 °C   (c) −62 °C   (d) 62 °C   (e) 61 °C

28 If I quadruple a number (*n*) and then take away 7, I get the same result as adding 23 to the same number (*n*). What is my original number (*n*)? (3)
   (a) 5 (b) 7 (c) 16 (d) 10 (e) 30

29 Albert has two different number cards.
   When he adds the numbers on the cards, the result is 26
   When he multiplies the two numbers, the result is 144
   What are the two numbers on the number cards? (2)
   (a) 13 and 13 (b) 12 and 14 (c) 16 and 10 (d) 17 and 11 (e) 18 and 8

30 When Lennox thinks of a number, divides it by 7 and then adds 24, his result is 30
   What was his original number? (3)
   (a) 54 (b) 42 (c) 8 (d) 49 (e) 61

31 The two-way table shows the colours and varieties of flowers in Sophia's bouquet.

	Roses	Tulips
Red	9	10
Yellow	6	7
Orange	7	5

   Without looking, Sophia picks one flower to give to her mum.
   Which of the following statements is true? (1)

   (a) Picking a red rose is more likely than picking a red tulip.
   (b) Picking an orange flower is more likely than picking a yellow flower.
   (c) Picking a rose and a tulip are equally likely.
   (d) Picking a red rose is less likely than picking a yellow rose.
   (e) Picking an orange rose and an orange tulip are equally likely.

32 Mohammed ran four 10 km races. His times are listed below.
   36 minutes 30 seconds   **34 minutes 10 seconds**
   38 minutes 45 seconds   **34 minutes 15 seconds**
   What is his mean time? (2)
   (a) 36 minutes 25 seconds (b) 35 minutes 40 seconds (c) 35 minutes 55 seconds (d) 35 minutes 25 seconds (e) 37 minutes 40 seconds

33 Jenny owns two square fields with a total area of 14 500 m². The first field has an area of 6400 m². What is the side length of the second field? (2)
   (a) 41 m (b) 121 m (c) 90 m (d) 65 m (e) 120 m

34 A class has x boys and y girls. There are 5 more boys than girls.
   What is the value of y? (1)
   (a) $y = 5x$ (b) $y = x + 5$ (c) $y = x - 5$ (d) $y = x$ (e) $y = 2x$

35 A 5p coin has a mass of 3.25 g.
   A bag of 5p coins has a mass of 650 g.
   How many 5p coins are in the bag? (1)
   (a) 20 (b) 50 (c) 100 (d) 200 (e) 250

Turn over to the next page

36 Harry made the shape shown from sticks. He wants to fill each face of the shape with cardboard. How many of each cardboard shape does Harry need? (1)

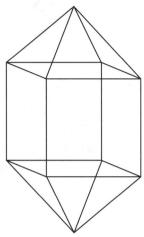

(a) 4 squares and 8 triangles
(b) 4 squares and 4 triangles
(c) 4 squares and 6 triangles
(d) 5 squares and 5 triangles
(e) 5 squares and 8 triangles

37 Paula enjoys running. For every 5 half-marathons that she runs, she runs 3 marathons. How many half marathons has she run when she has run 15 marathons? (3)
(a) 15 (b) 12 (c) 20 (d) 25 (e) 9

38 What is 78 × 386 + 386 × 22? (2)
(a) 8492 (b) 30 108 (c) 38 600 (d) 77 200 (e) 98 468

39 What is 6 × 4 − 5 × 3? (3)
(a) 57 (b) 9 (c) 39 (d) 66 (e) −18

40 Mr Thrower's allotment is 12 metres long. He digs 4 sections, each g m long. What length of the allotment does he have he left to dig? (1)
(a) $12 - g$ (b) $12g$ (c) $12 - 4$ (d) $g - 12$ (e) $12 - 4g$

41 The chart shows the height of some of the world's tallest buildings. Which two buildings have a difference of 39 metres in height? (1)

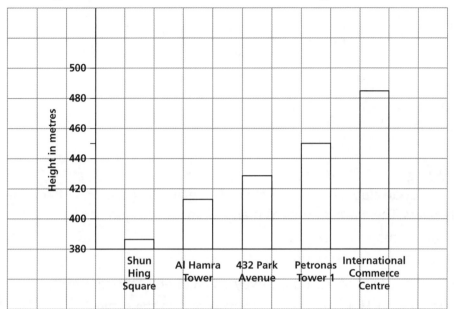

(a) 432 Park Avenue and Petronas Tower 1
(b) International Commerce Centre and Petronas Tower 1
(c) Shun Hing Square and 432 Park Avenue
(d) Al Hamra Tower and Petronas Tower 1
(e) 432 Park Avenue and International Commerce Centre

42 The length of one side of the shape shown is $3a + 1$
What is the perimeter of the shape? (1)

(a) $3a + 7$ (b) $3a + 6$ (c) $9a + 6$ (d) $18a + 1$ (e) $18a + 6$

43 The numbers of emails received by a small business each day over a two-week period are listed below.

14, 12, 22, 21, 33, 23, 12, 21, 35, 21, 27, 30, 21, 23

What is the mode? (1)

(a) 21 (b) 23 (c) 21.5 (d) 22.5 (e) 35

Turn over to the next page

44 The letters to spell 'MISSISSIPPI RIVER' are written on cards. They are shuffled and placed face down. What is the probability of picking a card with an S or a P? (2)

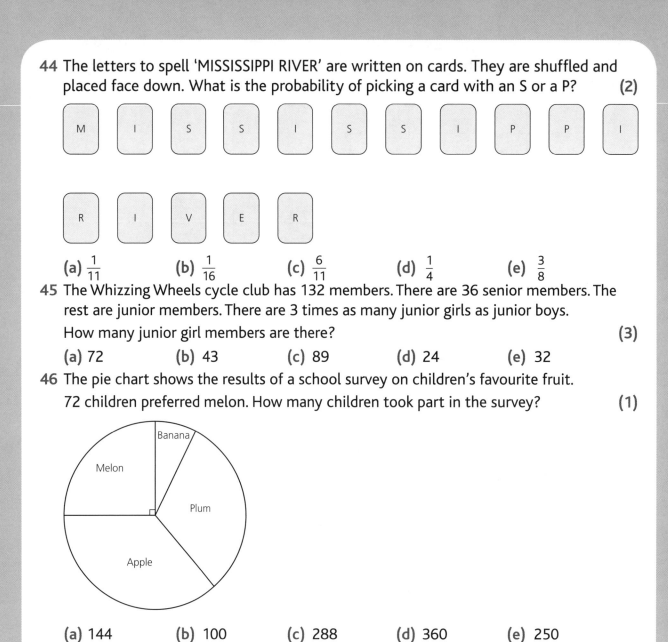

(a) $\frac{1}{11}$ (b) $\frac{1}{16}$ (c) $\frac{6}{11}$ (d) $\frac{1}{4}$ (e) $\frac{3}{8}$

45 The Whizzing Wheels cycle club has 132 members. There are 36 senior members. The rest are junior members. There are 3 times as many junior girls as junior boys. How many junior girl members are there? (3)

(a) 72 (b) 43 (c) 89 (d) 24 (e) 32

46 The pie chart shows the results of a school survey on children's favourite fruit. 72 children preferred melon. How many children took part in the survey? (1)

(a) 144 (b) 100 (c) 288 (d) 360 (e) 250

47 Which of these nets will form a triangular prism? (1)

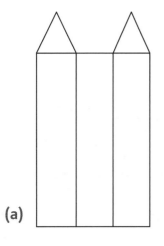 (a)

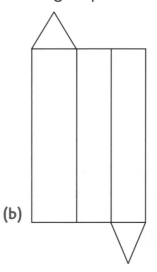 (b)

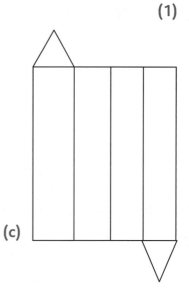 (c)

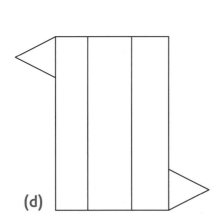 (d)

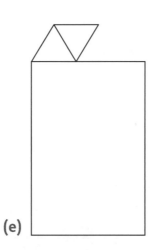 (e)

48 The line graph shows the amount of money Tobias would raise by running laps of a running track.
How much money would Tobias raise if he ran 45 laps? (1)

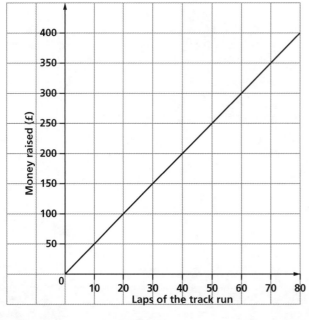

(a) £45 (b) £90 (c) £200 (d) £225 (e) £125

Turn over to the next page

49 Triangle A is an isosceles triangle. What is the size of angle x? (4)

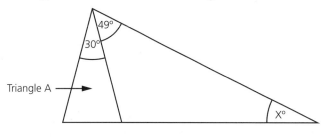

Not drawn accurately

(a) 30°   (b) 64°   (c) 26°   (d) 32°   (e) 52°

50 Patrick is counting stars in the night sky. He counts the number of stars in the sky each hour from dusk until dawn. His results are listed below.

3   8   38   68   196   200   180   210   142   81   25   3

What is the median number of stars he counts in an hour? (2)

(a) 3   (b) 96   (c) 74.5   (d) 81   (e) 190

Record your results

Score ☐ / 96   Time ☐:☐

Answers

Guidance on the breakdown of marks is given in brackets within the questions.

Paper 1

1. (e) 213
2. (a) 393
3. (b) 152
4. (e) $\frac{1}{4}$ $\frac{1}{2} = \frac{2}{4}$ (1) $\frac{3}{4} - \frac{2}{4} = \frac{1}{4}$ (1)
5. (a) 32 $4\overline{)128}$ (short division method)
6. (e) 144   90 + 54 = 144
7. (d) 130   58 + 67 = 125 (1)   125 rounds to 130 (1)
8. (a) 272   945 − 673
9. (b) 35   45 ÷ 9 = 5 (1)   5 × 7 (1)
10. (b) 90   Approximately 6 circles down and 16 across, 6 × 16 = 96
11. (b) 7   28 ÷ 4
12. (a) £3   7 + 1 = 8 (1)   £24 ÷ 8 = £3 (1)
13. (e) $\frac{3}{10}$ Divide the numerator and denominator by 2
14. (b) 0.75
15. (a) 7.5 m   100 cm = 1 m, 750 ÷ 100
16. (c) 360 cm   Width is 120 cm ÷ 2 = 60 cm (1)
    Perimeter = (120 cm × 2) + (60 cm × 2) = 240 cm + 120 cm (1)   = 360 cm (1)
17. (b) 51 runs   49 + 76 + 52 + 27 = 204 (1)   204 ÷ 4 (1)
18. (d) 3 $a = 12 − 9$ (1)   12 − 9 = 3 (1)
19. (e) 9 $b = 2 + 7$ (1)   2 + 7 = 9 (1)
20. (e) Sides are different lengths

Paper 2

1. (a) (i) £15.65   £1.85 × 5 = £9.25 (1)   £1.60 × 4 = £6.40 (1)   = £9.25 + £6.40 = £15.65 (1)
   (b) (iii) 5 coins   Coins of £2, £2, 20p, 10p, 5p
   (c) (v) £7.55   £6.40 ÷ 2 = £3.20 (1)   £15.65 − £3.20 = £12.45 (1)   £20.00 − £12.45 = £7.55 (1)
   (d) (ii) £8   £7.55 = 755p, 755p rounds to 800p = £8.00
   (e) (ii) £1, 50p, 10p, 2p, 1p
2. (a) (v) 3.48 m   3.65 m + 3.45 m + 3.25 m + 3.57 m = 13.92 m (1)   13.92 m ÷ 4 (1)
   (b) (iv) 0.4 m   Identify the smallest measurement (3.25 m), identify the largest measurement (3.65 m),
             3.65 m − 3.25 m
3. (a) (v) 16   Michelle receives 8 sweets from her 2 parts, so each part is 4 sweets (1)
            Peter gets 4 × 4 = 16 (1)
   (b) (iii) 36   4 + 3 + 2 = 9 (1)   9 × 4 (1)
4. (c)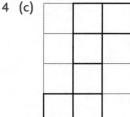

5. (a) (i) 40 m   14 m + 6 m + 8 m + 3 m + 6 m + 3 m

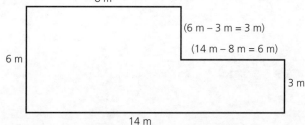

   (b) (iii) 66 m²   8 m × 6 m = 48 m² (1)   6 m × 3 m = 18 m² (1)   48 m² + 18 m² = 66 m² (1)
   (c) (v) £554.40   66 m² × £8.40

Mathematics Practice Papers 1 published by Galore Park

6 (a) (ii) 7   26 − 13 = 13 (1)   20 − 13 = 7 (1)
  (b) (iv) 5   2 × 5 = 10 (1)   50 ÷ 10 = 5 (1)
7 (a) (iii) 32   Add the tally marks 5 + 5 + 5 + 5 + 5 + 5 + 2
  (b) (v) 200   Calculate the total for each type of travel (1)   57 + 68 + 32 + 32 + 11 = 200 (1)
  (c) (iii) 36   The number of children who came by car is 68. The number who came by bus is 32.
                      68 (1) − 32 (1) = 36 (1)
  (d) (i) $\frac{1}{2}$ 57 (walking) + 32 (bicycle) + 11 (scooter) = 100 (1) $\frac{100}{200}$ (1)   = $\frac{1}{2}$ (1)
8 (b) (1, 2) (1), (−1, 0) (1)

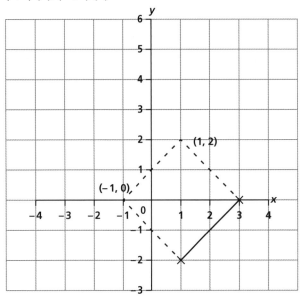

Paper 3

1 1059   M = 1000, L = 50, IX = 9
2 9   27 − 18
3 3, 3 (1) for correct hundreds digit, (1) for correct units digit

	3	8	9
+	3	5	3
	7	4	2
		1	1

4 23
5 3762   1000 × 3 = 3000, 200 × 3 = 600, ...   3000 + 600 + 150 + 12 = 3762
6 2926   5272 − 2346
7 132   12 × 11 = 132
8 30   2 × 3 × 5
9 £306   Reduction £360 × 15 ÷ 100 = £54 (1)   New price £360 − £54 = £306 (1)
10 $\frac{2}{5}$ $\frac{10}{25}$ (1)   = $\frac{2}{5}$ (1)
11 $y = 6$ $3y = 43 − 25$ (1) $3y = 18$ (1) $y = 6$ (1)
12 18.4 cm   Opposite sides are of equal length (1)   6.4 cm + 6.4 cm + 2.8 cm + 2.8 cm = 18.4 cm (1)

13 25 cm²   21 whole squares + 8 half squares

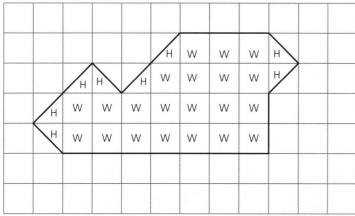

14 9.97 kg   3.24 kg + 0.81 kg + 4.22 kg + 1.7 kg
15 48   64 ÷ 4 = 16 (1)   16 × 3 = 48 (1)
16 35   3y + 8 = (3 × 9) + 8 = 35
17 9.88 s   (9.81 s + 9.85 s + 9.91 s + 9.83 s + 10.0 s) = 49.40 s (1)   49.40 s ÷ 5 = 9.88 s (1)
18 45 km   5 km × 9
19 200°   90° + 50° + 20° = 160° (1)   360° − 160° = 200° (1)
20 400 cm or 4 m   40 cm × 10

Paper 4

1 (a) 48 m   12 m + 12 m + 12 m + 12 m
   (b) 44 m²   12 m × 12 m = 144 m² (1)   10 m × 10 m = 100 m² (1)   144 m² − 100 m² = 44 m² (1)
   (c) 2304 tiles   0.25 m × 0.25 m = 0.0625 m² (1)   144 m² ÷ 0.0625 m² (1) or alternative method
   (d) £230.40   2304 tiles × 10p
   (e) £70.40   100 m² ÷ 0.0625 m² = 1600 tiles (1)   1600 tiles × 10p = £160 (1)
                      £230.40 − £160 = £70.40 (1) or alternative method
2 (a) 68 516   82 000 − 13 484
   (b) 3371   25% of 13 484 = 13 484 ÷ 4
   (c) 6000   8000 ÷ 4 = 2000 (1)   2000 × 3 (1)
3 One million, forty thousand, seven hundred and three
4 £6.78   £6.87   £7.86   £8.67   £8.76
5 (a) 27
   (b) 21   48 − 27
   (c) 105   38 + 27 + 21 + 19
   (d) 65   38 + 27
   (e) 59   38 + 21
6 (a) $3(x − 5) = 12$ (1) for $x − 5$ (1) for 3 × bracket
   (b) 9   12 ÷ 3 = 4 (1)   4 + 5 = 9 (1)

7

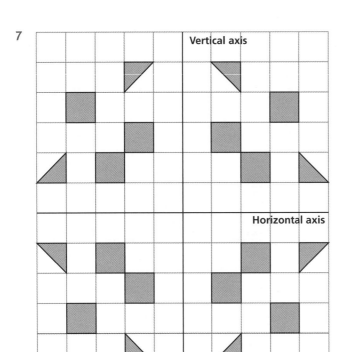

(2)

8 (a)

(b) 920 m   230 m × 4

Paper 5

1 15 cm   49 cm $\frac{1}{2}$m   51 cm   98 cm   1 m   Any 3 measurements in order (1), all 6 in the correct order (2)
2 1460
3 529
4 5   5 is divisible only by 1 and itself
5 (1, 24)   (2, 12)   (3, 8)   (4, 6)
6 17   (3 × 3) = 9 (1)   (2 × 2 × 2) = 8 (1)   9 + 8 = 17 (1)
7 240
8 $\frac{1}{3}$ $\frac{62}{186}$ (1)   cancels to $\frac{1}{3}$ (1)
9 $\frac{4}{10}$ or $\frac{2}{5}$ $\frac{1}{2} \times \frac{4}{5} = \frac{4}{10} (= \frac{2}{5})$
10 27   45 ÷ 5 = 9 (1)   9 × 3 = 27 (1)
11 0.291
12 65 miles   78 miles ÷ 6 = 13 miles (1)   13 miles × 5 (1)
13 $\frac{4}{12}$ or $\frac{1}{3}$
14 8   3 × 7 + y = 29, 21 + y = 29 (1)   y = 29 − 21 = 8 (1)
15 39   3, 7, 11, 15, 19, 23, 27, 31, 35, 39
16 1405 m   281 m × 5
17 4.25 km   4250 m ÷ 1000
18 20 cm   8 × 2.5 cm
19 38°   90° + 52°= 142° (1)   180° − 142° (1)
20 8   The rugby section is 90°, i.e. one-quarter of the pie chart (1)   Total children is 32, so 32 ÷ 4 (1)

46

Paper 6

1. (a) 740 km
   (b) 110 km   To Rennes
   (c) 275 km   Marseille to Rennes 1050 km (1)   Marseille to Paris 775 km (1)   1050 km − 775 km = 275 km (1)

2. (a) 39 km $3 \times 12 = 36$ km (1)   20 minutes is $\frac{1}{3}$ of an hour, $\frac{1}{3}$ of 9 km = 9 km ÷ 3 = 3 km (1)   36 km + 3 km = 39 km (1)
   (b) 11.7 km/h $39 \div 3\frac{1}{3}$ (1) $= 39 \times \frac{3}{10}$ (1) $= \frac{117}{10}$ (1)

3. (a) 70%   66 + 78 + 90 + 60 + 62 + 80 + 68 + 56 = 560 (1)   560 ÷ 8 = 70% (1)
   (b) 12   Natasha: 80% = $\frac{80}{100} = \frac{40}{50}$ (1)   Luke: 56% = $\frac{56}{100} = \frac{28}{50}$ (1)   40 − 28 = 12 (1)
   (c) 17 marks   Ellie: 90% = $\frac{90}{100} = \frac{45}{50}$ (1)   Luke scored $\frac{28}{50}$ (above)   45 − 28 = 17 (1)

4. (a) 50 litres   450 km ÷ 9 km/litre
   (b) £55   50 litres × £1.10

5. (a) 125   5 cm × 5 cm × 5 cm
   (b) 27   98 cubes have a least 1 face showing (1)   125 − 98 = 27 (1)
   (c) 78.4%   98 cubes have a least 1 face showing, 98 ÷ 125

6. (a) £98.62   Ties: £8.95 × 7 = £62.65 (1)   Shirts: £11.99 × 3 = £35.97 (1)   £62.65 + £35.97 = £98.62 (1)
   (b) £87.57   New total £98.62 + £8.95 = £107.57 (1)   £107.57 > £100 so £107.57 − £20 = £87.57 (1)

7. (a) 1 hour 4 minutes   11:38 − 10:34
   (b) 10:34   Arrives at Gatwick at 11:08
   (c) 13:01   12:52 + 9 minutes
   (d) 1 hour 36 minutes   13:07 train
   (e) 445 000   981 000 − 536 000

8. (a) 10 trees   Each tree symbol stands for 2 trees
   (b) 55 trees   Oak: 8 trees, Ash: 10 trees, Horse chestnut: 6 trees, Hazel: 9 trees, Yew: 3 trees, Sycamore: 9 trees, Hawthorn: 10 trees (1)   8 + 10 + 6 + 9 + 3 + 9 + 10 = 55 (1)
   (c) 838 m   (Oak) 8 × 40 m = 320 m (1)   (Ash) 10 × 35 m = 350 m (1)   (Horse chestnut) 6 × 28 m = 168 m (1)   320 m + 350 m + 168 m = 838 m (1)
   (d) 11 trees   6 horse chestnut trees × 11 = 66 trees in total (1)   66 − 55 (original trees) = 11 (1)
   (e) 33 trees   2 and 3 = 5, 55 trees ÷ 5 = 11 trees (1)   11 trees × 3 = 33 (1)

9. (a)

Number of goals	Tally of goals scored	Frequency
2	I	1
3	III	3
4	III	3
5	IIII	4
6	IIII I	5
7	IIII	4
8	II	2
9	II	2

   (b) 5.625 goals   Total goals 135 (1)   135 ÷ 24 (1)
   (c) 6 goals   modal score is the most common score
   (d) 6 goals   the median is the middle score of an ordered set of scores (1)
                              2, 3, 3, 3, 4, 4, 4, 5, 5, 5, 5, **6, 6**, 6, 6, 6, 7, 7, 7, 7, 8, 8, 9, 9 (1)
   (e) 15 goals   25 matches × 6 goals = 150 goals (1)   135 goals in 24 matches, 150 − 135 = 15 (1)

10. £989   £8901 ÷ 9

11 (a) (1, 3) (3, 1) (6, 4) (4, 6)
   (b) (1) for stone reflected in line A
   (c) (1) for each 'stone' reflected in line B

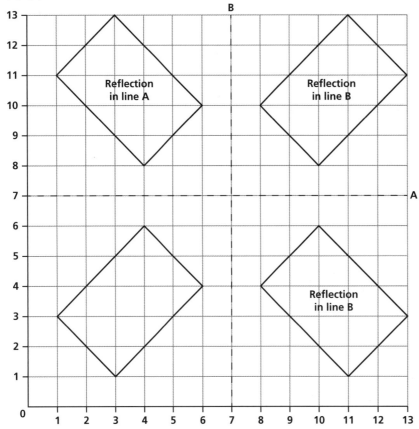

12 6   27 × 2 = 54 (1)   54 ÷ 9 (1)
13 (a) 1200   48 × 25
   (b) 2400 kg   1200 × 2 kg
14 20.5 km   3936 km ÷ 24 = 164 km (1) 164 km ÷ 8 = 20.5 km (1)
15 19 straws $2x + 1$, where x is the number of patterns (1)   (2 × 9) + 1 (1)
16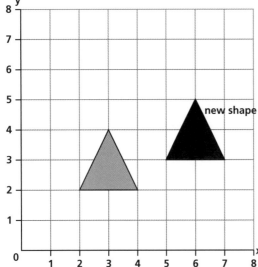

Paper 7

1. >   6 + 115 ___ 121 − 3 (1)   121 > 118 (1)
2. 47, 95, 191   (1) for each term   (Double the previous number and add 1)
3. 12   Factors of 24 – 1, 2, 3, 4, 6, 8, **12**, 24 (1)   Factors of 36 – 1, 2, 3, 4, 6, **12**, 18, 36 (1)
4. 2, 3 and 7   1, **2**, **3**, 6, **7**, 14, 21, 42
5. 2500 mm $2\frac{1}{2}$ m × 1000
6. 67 666   76 776   666 766   767 667   776 677
7. 498 $\frac{1}{3}$ of the book is 166 pages, so 166 × 3
8. 122.5 cm³   7 cm × 5 cm × 3.5 cm
9. 36, 44, 48, 56, 64 and 68 shaded (1)   36, 48, 54 and 66 circled (1)
10. 5, 6 and 7   5 + 6 + 7 = 18 or 18 ÷ 3 = 6 (middle of consecutive numbers) (1)   5 × 6 × 7 = 210 (1)
11. $\frac{2}{5}$ $\frac{4}{5} \times \frac{1}{2} = \frac{4}{10}$ (1)   cancels to $\frac{2}{5}$ (1)
12. $\frac{1}{15}$ $\frac{1}{5} \times \frac{1}{3}$ Multiply the numerators and the denominators
13. $2\frac{4}{6}$ or $2\frac{2}{3}$ Take away whole number from whole number: 4 − 2 = 2 (1)
    Take away fraction from fraction $\frac{5}{6} - \frac{1}{6} = \frac{4}{6}$ (1)
14. $5\frac{5}{6}$ 2 + 3 = 5 (1) $\frac{3}{6} + \frac{2}{6} = \frac{5}{6}$ (1)
15. £99   £660 × 15 (1) = 9900   9900 ÷ 100 (1) or alternative method
16. £445   3 + 4 + 5 = 12 (1)   1068 ÷ 12 = £89 (1)   £89 × 5 (1)
17. 23   Two-digit prime numbers < 30: 11, 13, 17, 19, 23, 29 (1)   2 + 3 = 5 (1)
18. 1000   20 cm ÷ 2 cm = 10 (1)   10 × 10 × 10 (1) or alternative method
19. Hexagonal prism   8 faces   18 edges   12 vertices   (1) for each correct answer
20. 37.68 cm   3.14 × 12
21. $n + 24$ or $\frac{n}{4} + 6$
22. 1038   346 × 3 = (300 × 3) + (40 × 3) + (6 × 3) = 900 + 120 + 18
23. 1456   Use bus stop method
24. 3.575 kg   (2.9 kg + 3.7 kg + 3.1 kg + 3.9 kg + 4.2 kg + 3.6 kg + 3.1 kg + 4.1 kg) = 28.6 kg (1)   28.6 kg ÷ 8 = 3.575 kg (1)
25. $a = 2$ 5 + $2a^2$ = 13 (1) $2a^2 = 8$ (1) $a^2 = 4$ (1) $a = 2$ (1) or 20 + $8a^2$ = 52 (1)
    $8a^2 = 32$ (1) $a^2 = 4$ (1) $a = 2$ (1)
26. 4 : 3   120 apples : 90 oranges, 4 apples : 3 oranges
27. £8, £12   2 + 3 = 5 (1)   £20 ÷ 5 = £4 (1)   £4 × 2 (1)   and   £4 × 3 (1)
28.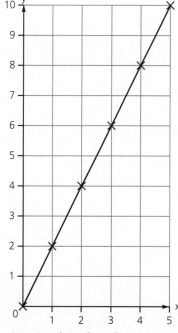
    y is twice the value of x
29. 9 inches   2.5 cm : 1 inch, 22.5 cm ÷ 2.5 cm
30. 2 hours 14 minutes   134 minutes − 120 minutes (2 hours) = 14 minutes

Mathematics Practice Papers 1 published by Galore Park

Paper 8

1 (a)

	Swimming club	Running club	Rugby club	Total
Mr Richards' class	12	8	8	28
Mrs Flynn's class	10	13	7	30
Mrs Thomas's class	5	13	9	27
Total	27	34	24	85

(1) for any 2 correct, (2) for any 4 correct, (3) for any 6 correct, (4) for all 7 correct
(b) 33.3%   10 ÷ 30 (1)   0.333 × 100 (1)
(c) 9   1 + 2 = 3 (1)   27 ÷ 3 = 9 (1)
(d) 3 : 2 : 2   12 : 8 : 8
(e) 17   10% of 85 is 8.5 (1)   8.5 × 2 = 17 (1) *or alternative method*

2 (a)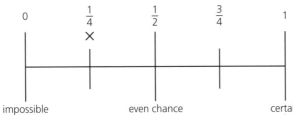

2 square numbers (1 and 4) from a total of 8 cards $\frac{2}{8} = \frac{1}{4}$ of the cards show a square number

(b)

6 even numbers (2, 4, 6, 8, 12 and 24) $\frac{6}{8} = \frac{3}{4}$ of the cards show an even number

(c)

8 factors of 24 (1, 2, 3, 4, 6, 8, 12 and 24) $\frac{8}{8} = 1$, all the cards show a factor of 24

3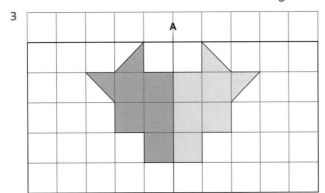

4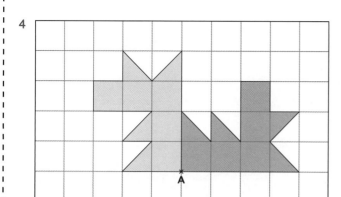

5 (a) 1 Arjun
   (b) 2 Joe and Phoebe
   (c) 3 Anne, Darren and Arjun
   (d) 1 Anne
6 (−4,−2) (2,−2) and (−1, 2)   (1) for each pair of co-ordinate values

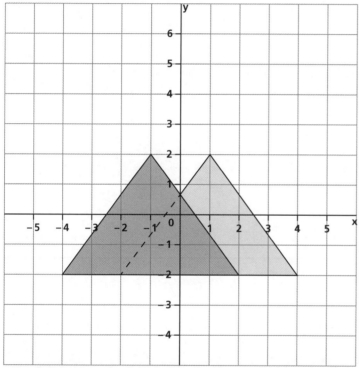

7 (a) 1400 B   £350 × 4
   (b) £133.75   120 B + 85 B + 100 B + 230 B = 535 B (1)   535 B ÷ 4 (1)
   (c) 60 B   120 × 4 = 480 B (1)   120 × 3.5 = 420 B (1)   480 B − 420 B (1)
   (d) £50   150 B ÷ 3
   (e) UK, by 6 B   UK price in Blings: £28.50 × 4 = 114 B (1)   120 B − 114 B = 6 B (1)
8 (a) 35 000   100 × 350
   (b) 36 750   5 × 350 = 1750 (1)   35 000 + 1750 = 36 750 (1)
   (c) 35 days   100 days + 5 days = 105 days (1)   105 days ÷ 3 (1)
9 (a) 10.5 m   1750 mm × 6 = 10 500 mm (1)   10 500 mm ÷ 1000 = 10.5 m (1)
   (b) 28.44 m   4740 mm × 6 = 28 440 mm (1)   28 440 mm ÷ 10 000 = 28.44 m (1)
10 (a) £2160   £17 248 ÷ 8 = £2156 (1)   Round to the nearest £10 (1)
   (b) £83   £2156 ÷ 26 = £82.92 (1)   Round to the nearest £ (1)
11 (a) 8 cherries $a = 2b$, $2b = 8c$ (1), so $a = 8c$ (1)
   (b) 6 bananas $3a = 6b$ (1), so kiwi = $6b$ (1)
12 364   Add the previous term (1) to the difference, which increases by a factor of 3 each time (1)

Mathematics Practice Papers 1 published by Galore Park

Paper 9

1. (d) 7   14 + 14 + 14 + 14 + 14 + 14 = 84 (1)   84 ÷ 12 = 7 (1)
2. (c) $\frac{1}{6}$ $\frac{2}{12}$ (1)   = $\frac{1}{6}$ (1)
3. (d) 9.6 = $9\frac{3}{5}$
4. (b) £418 058
5. (a) 40   70 − 30 (1)   = 40 (1)
6. (d)
7. (c) 23.35
8. (e) 15 232   16 is $\frac{1}{2}$ of 32 (1)   30 464 ÷ 2 = 15 232 (1)
9. (a) 56   2 litres × 6 = 12 litres (1)   12 litres + 2 litres = 14 litres (1)   14 litres = 14 000 ml (1)
   14 000 ml ÷ 250 ml = 56 (1)
10. (b) 1.3
11. (e) a parallelogram

12. (a) 31   32 + 43 + 37 + 33 + 26 + 26 + 20 = 217 (1)   217 ÷ 7 = 31 (1)
13. (c) (3, 7)

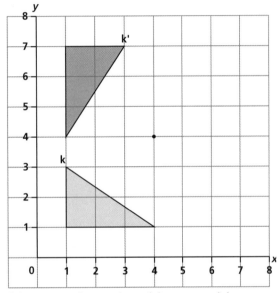

14. (d) $4W − 1$ $(8W − 2) ÷ 2$ (2) $4W − 1$ (1)
15. (c) (5, 5)

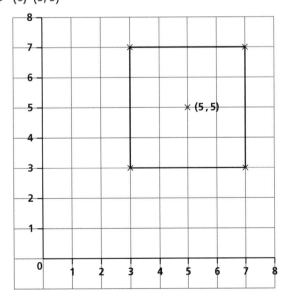

(1) for correct *x*-co-ordinate (1) for correct *y*-co-ordinate

16 (c) (2, −3)

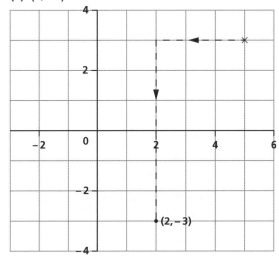

(1) for correct x-co-ordinate (1) for correct y-co-ordinate

17 (d) 2, 2, 8   (1) for each digit   92 + 283 = 375
18 (b) 1320
19 (d) 15   2.5 × 2 × 3 (1)   = 15 (1)
20 (c) 12% $\frac{3}{25}$ (1)   = $\frac{12}{100}$ = 12 % (1)
21 (a) 468   Terms are in the 6 times table, numbers are multiples of 6 (1)   468 (1) is divisible by 6
22 (d) 184   (1) for each of the criteria satisfied
23 (c) £23.76   100p × 24 = 2400p (1)   2400p − 24p = 2376p (1)   = £23.76 (1)
24 (b) Richard   15 months = 1 years 3 months (1)   21/07/14 to 15/10/15 is closest (1)
25 (a) 483   23 + 25 = 48 (1)   968 − 48 = 920 (1)   920 students − 46 = 874 (1)   874 ÷ 2 = 437 girls (1)
            437 + 46 = 483 boys (1)
26 (b) 07:15   Arrive at Hopesfield bus station by 09:00 − 5 minutes = 08:55 (1)
            Arrive 08:49 so 07:15 from Marston (1)
27 (c) −62 °C   The scale goes up in 2 °C steps
28 (d) 10 $4n − 7 = n + 23$ (1) $3n = 30$ (1) $n = 10$ (1)
29 (e) 18 and 8 (1) for each of the criteria satisfied
30 (b) 42 $n ÷ 7 + 24 = 30$ (1) $n = (30 − 24) × 7$ (1) $n = 6 × 7 = 42$ (1)
31 (c) Picking a rose and a tulip are equally likely.
32 (c) 35 minutes 55 seconds   143 minutes 40 seconds (1)   143 minutes 40 seconds ÷ 4 = 35 minutes
                    55 seconds (1)
33 (c) 90 m   14 500 m² − 6400 m² = 8100 m² (1)   Square field, 90 m × 90 m (1)
34 (c) $y = x − 5$
35 (d) 200   650 g ÷ 3.25 g = 200
36 (a) 4 squares and 8 triangles (4 from each square-based pyramid)
37 (d) 25   5 half marathons to 3 marathons is 5 : 3 (1)   15 marathons ÷ 3 = 5 (1)
            5 half marathons × 5 = 25 half marathons (1)
38 (c) 38 600   78 × 386 + 386 × 22 = (78 + 22 = 100) (1)   100 × 386 = 38 600 (1)
39 (b) 9   24 (1) − 15 (1) = 9 (1). Use BIDMAS
40 (e) $12 − 4g$

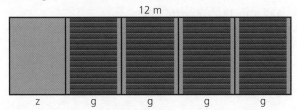

41 (d) Al Hamra Tower and Petronas Tower 1   452 m − 413 m = 39 m
42 (e) $18a + 6$ $(3a + 1) × 6 = 18a + 6$
43 (a) 21   The mode is the most common.
44 (e) $\frac{3}{8}$ $\frac{6}{16}$ (1) = $\frac{3}{8}$ (1)
45 (a) 72   Total members − senior members = junior members: 132 − 36 = 96 (1)   96 ÷ 4 = 24 (1)
            24 × 3 = 72 (1)
46 (c) 288   Melon is favourite for $\frac{1}{4}$ of the children, 72 × 4 = 288

Mathematics Practice Papers 1 published by Galore Park

47 (b)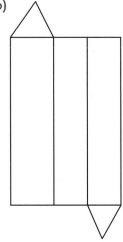

48 (d) £225
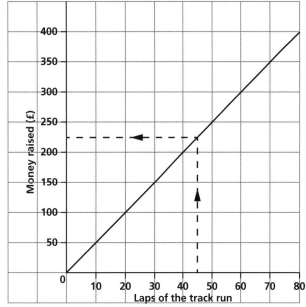

49 (c) 26°
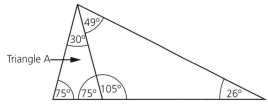

Triangle A is isosceles with base angles of 75° [180° − 30° = 150° (1)   150° ÷ 2 = 75° (1)]
Angle next to 75° is 180° − 75° = 105° (1)   180° in a triangle, so x = 180° − (49° + 105°) = 26° (1) or other appropriate method.

50 (c) 74.5   Correctly ordered data [3 3 8 25 38 68 81 142 180 196 200 210] (1)
Mean of 2 middle values (68 + 81) ÷ 2 = 74.5 (1)